María José Luciáñez Sánchez
Enrique Rubio Gómez
Marta Vega Manchón

The effect of fire on the forest

María José Luciáñez Sánchez
Enrique Rubio Gómez
Marta Vega Manchón

The effect of fire on the forest

Springtails as bioindicators of the degradation of the soil environment.

ScienciaScripts

Imprint

Cover image: www.ingimage.com

This book is a translation from the original published under ISBN 978-613-9-40485-8.

Publisher:
Sciencia Scripts
is a trademark of
Dodo Books Indian Ocean Ltd. and OmniScriptum S.R.L publishing group

120 High Road, East Finchley, London, N2 9ED, United Kingdom
Str. Armeneasca 28/1, office 1, Chisinau MD-2012, Republic of Moldova, Europe
Managing Directors: Ieva Konstantinova, Victoria Ursu
info@omniscriptum.com

Printed at: see last page
ISBN: 978-620-8-50442-7

THE EFFECT OF FIRE ON THE FOREST: SPRINGTAILS AS BIO-INDICATORS

THE EFFECT OF FIRE ON THE FOREST FLOOR: SPRINGTAILS AS BIOINDICATORS OF SOIL DEGRADATION

MARÍA JOSÉ LUCIÁÑEZ SÁNCHEZ

ENRIQUE RUBIO GÓMEZ AND MARTA VEGA MANCHÓN

INDEX

SUMMARY

In this work, a faunistic and ecological study was carried out in a pine forest of Pinus nigra Arn. in Fuentenava de Jábaga, in the Serranía de Cuenca, with the aim of clarifying the impact of forest fires on the composition of the mesofauna and on springtail communities. During the autumn season, samples of leaf litter, surface soil and deep soil were collected from different parts of the natural forest, the area affected by the fire and the transition zone between the two. Analysis of abundance, distribution and diversity indices revealed that there is a lower abundance of springtail populations in the burned forest, although it is the area with the highest diversity value.

Through the different statistical analyses carried out, it was possible to identify the springtail species associated with each of the environments (natural, fire or transition zone), as the species are influenced by the edaphic characteristics of each type of soil, which confirms their role as bioindicators of edaphic ecosystems. Euedaphic species, such as Mesaphorura macrochaeta, are highly adapted to living in fire-disturbed environments, so they are very abundant in the burned soil. On the other hand, more numerous communities are observed in the transition or edge zone, especially of Tetracanthella pilosa, a hemi-daphic species with a holarctic distribution which, due to the laying of dormant eggs that hatch at high temperatures, is particularly relevant in this ecosystem.

KEY WORDS: *Collembola, Pinus nigra, fire, soil bioindicators, Serranía de Cuenca.*

1. INTRODUCTION

Scientific knowledge and the different concepts related to it are evolving at the same time. Although it is an old and therefore perhaps well-established concept, the definition of soil has never ceased to be controversial. It could be considered as a natural body, differentiated into horizons of generally unconsolidated mineral and organic constituents, of variable depth and differing from the original parent rock in morphology, constitution, composition, physical and chemical properties and biological characteristics (JOFFE, 1936). Soil is a limited and non-renewable natural resource that performs multiple ecosystem or environmental services, including the regulation of biogeochemical cycles, carbon fixation and water storage and filtration (BURBANO-ORJUELA, 2016). In addition, soil is home to a high density of species, with numerous animal phyla and microflora species, making it the most biodiverse natural ecosystem in any region (HÅGVAR, 1998).

Soil macrofauna (ants, spiders, pseudoscorpions...) have a significant impact on decomposition and nutrient mineralisation processes. Mesofauna, which includes arthropods between 0.2 and 2 mm in size, such as mites or springtails, are even more abundant and diverse (MURILLO-CUEVAS et al., 2019). Some groups are sensitive to natural or human-induced disturbances, which makes them bioindicators of soil quality and fertility (GARCÍA-ÁLVAREZ & BELLO, 2004). It is essential to investigate the disturbances that may affect these edaphic communities in order to understand the dynamics and state of forest ecosystems. In this work we aim to study the effects of forest fires on the soil and how these disturbances affect the communities of organisms and their functioning.

Mediterranean ecosystems have developed a high adaptation to forest fires, however, the frequency of forest fires has increased significantly. Nowadays, depopulation and abandonment have led to an increase in vegetation cover and an increase in uncontrolled and high intensity forest fires (BODÍ et al., 2012).

Forest fires have a strong impact on biogeochemical cycles, vegetation, soil, fauna, hydrological and geomorphological processes, water quality and even the composition of the atmosphere (BODÍ et al., 2012). These events destroy litter and desiccate the upper soil layers, leading to changes in the availability of food, water, temperature and pH. The organic matter remaining after the fire is often highly decomposed and provides a less rich substrate for decomposer organisms compared to unburned soil and litter (SUHADI et al., 2012).

The "fire regime" brings together a series of characteristics that define these events (frequency, intensity, seasonality, area affected and type of spread), and according to ÚBEDA et al. (2021), since the 1960s this regime has changed, increasing the intensity and area affected by Large Forest Fires, in addition to the fact that they are occurring more and more frequently outside the summer months. Fires not only generate losses of organic matter, but also alter soil properties, increasing pH and salinity, favouring decomposer microorganisms and altering biogeochemical cycles (phosphorus increases, nitrogen is lost and the quantity and quality of organic carbon is altered), as well as eroding the soil and even causing desertification (MATAIX-SOLERA & GUERRERO, 2007).

However, fires also provide benefits by allowing the regeneration of ecosystems, light penetration into the forest, and by releasing nutrients that would otherwise remain immobilised. These factors determine the composition of post-fire soil communities.

Springtails as bio-indicators

Springtails (Collembola) are a group of terrestrial hexapods, usually edaphic, that form a taxon phylogenetically close to insects. According to CIPOLA et al. (2018), the characteristics that separate this group from insects are their endognathous character and the presence of three singular abdominal appendages: a ventral tube or colophore, a tenaculum or retinaculum and the furca, although the latter two are lost secondarily in some taxa. The class has four orders: Poduromorpha, Entomobryomorpha, Symphypleona and Neelipleona.Within the edaphic faunal communities, they are one of the most relevant groups. In general, they feed on fungi and decomposing plant matter, but some of them are carnivores (they eat nematodes, tardigrades, other springtails, etc.) and very few feed on algae and living plants. They are highly important in edaphic food chains, contributing excretions, excretions, secretions or their remains after death to the environment, and in turn, they can serve as food for other arthropods such as ants, beetles or mites (ARANGO-GALVÁN et al., 2009). However, they only represent around 1-5% of the biomass in temperate systems and 10% in arctic areas, although in early stages of succession they can reach 33% (ARANGO-GALVÁN et al., 2009).

They are essential organisms for the good condition of the soil due to their important role in the decomposition of organic matter, which favours the establishment of microflora and facilitates the dispersion and activity of bacteria and fungi (BELLINGER et al., 2003), thus controlling the populations of microorganisms (especially fungi). They are therefore of great importance for assessing the state of a soil ecosystem and thus the effects of forest fires on the soil, as they also have a remarkable capacity for adaptation and can reach higher population levels than before the fire approximately three years later (LOPES & GAMA, 1994; DI CASTRI & VITALI DI CASTRI, 1981). They are among the first organisms to react to changes, being animals of high bioindicator value for soil environments (LUCIÁÑEZ & INIESTO, 2006).

Springtails, together with other groups of animals such as mites, symphylans, pauropods, diplurids, proturids, psocoptera, tisanoptera and encystreids, make up the soil mesofauna (animals 0.2-2 mm in diameter) and are essential in many of the processes occurring in the soil environment (BERUDE et al., 2015, SANJUAN et al., 2022). According to the morphoecological categories of GISIN (1943) (according to their anatomy and the environment in which they live associated with, along a vertical gradient), springtails can be characterised as atmobiose, hemiedaphic and euedaphic. Atmobiose are typical of the soil surface and vegetation, possessing long antennae and furcae, and developed pigmentation and vision. Hemiedaphids live in the leaf litter and top few centimetres of soil, and have more or less developed pigmentation and moderately long appendages. Euedaphids reside in deeper soil layers, and have reduced or absent colouring, eyes and appendages. Those with an epigean way of life (atmobios and hemiedaphic) can be: hygrophilous, closely associated with water; mesophilous, living in the superficial part of the leaf litter, or xerophilous, with intense pigmentation, living on mosses, lichens and bark (ARBEA & BLASCO-ZUMETA, 2001).

The great bioindicator value of this faunal group is the determinant of the subject of this text, which aims to contribute to the evaluation of the state of a soil, and therefore of an ecosystem, by studying the variations of springtail communities in a burnt soil. The study is part of a wider research project that aims to analyse and understand the ecology and behaviour of springtails in burnt soils using as a sampling point a forest of Pinus nigra Arn. in Fuentenava de Jábaga (Serranía de Cuenca), after a fire that took place in August 1991. The starting point is the analysis of the biogeographical and ecological characteristics of the species found, and from there and by means of numerical analysis, to contribute to the knowledge of the edaphic ecosystems and their functioning.

The initial hypothesis of this study is that the ecosystems studied show significant differences in the presence and abundance of the different groups of edaphic fauna and, especially, in springtail species, depending on the type of soil sampled: burnt soil, natural soil and transitional "edge" soil between burnt and natural soil. Furthermore, it is expected that springtail populations will recover over time.

2. OBJECTIVES

The main objective of this work is to contribute to the study of the impact of a forest fire on the composition of mesofauna and springtail populations. To this end, the specific objectives are as follows:

- To determine the abundance and distribution of soil fauna in a Pinus nigra pine forest in Fuentenava de Jábaga (Cuenca), in a natural state, in burnt soil, and in the transition zone between the two points, also studying different soil levels.

- Given the bioindicator value of edaphic springtails, the aim is to know the species that inhabit these natural and burnt soils, as well as in the transition zone, to study their distribution and population changes, and the factors that generate them.

- Identify possible groups of edaphic fauna and, above all, species of springtails that can serve as bio-indicators of a natural forest. of P. nigra, and a burnt forest. In this way they can become indicator species for soil quality and soil recovery after fire.

- To assess the effect of a fire on mesofauna, in order to better understand the impact of a forest fire, and to contribute to improving forest recovery.

3. DESCRIPTION OF THE STUDY AREA GEOGRAPHICAL SITUATION

The Serranía de Cuenca is located in the north-eastern region of the province of Cuenca and south of Toledo. It is not a high mountain area, but an area of rugged relief and intricate geological formations, covered by dense pine forests. It is located between the provinces of Cuenca, Guadalajara and Teruel, bordered to the north by the Sierra de Albarracín, to the east by the Serranía Celtibérica and to the west by the Alcarria. It is a region whose altitudes range from 800 metres at the bottom of some valleys to 1866 metres, with the Mogorrita peak, in the Sierra de Valdeminguete, being the highest point (DOMÍNGUEZ-FONTANA, 2011).

The study area consists of a pine forest located in the municipality of Fuentenava de Jábaga, bordered to the south by the N-400 road and to the east by the N-320 road, and about 13 km from the city of Cuenca (**figures 1 and 2**). The pine forest where the sampling was carried out is located in the Serranía de Cuenca, at an altitude of 943 m in Altos de Cabrejas, with UTM coordinates 30TWK6437.

GEOLOGY AND SOIL SCIENCE

In geological terms, the Serranía de Cuenca has a Triassic core, surrounded by Jurassic formations, which are overlain by Cretaceous and Tertiary strata. The landscape is dominated by moorland, the result of the karstic modelling. These paramounts are fragmented by intramountainous grooves forming valleys and gorges, i.e. fluviokarst erosion canyons with steep slopes and escarpments on massive Turonian dolomites (MONERO et al., 2010).

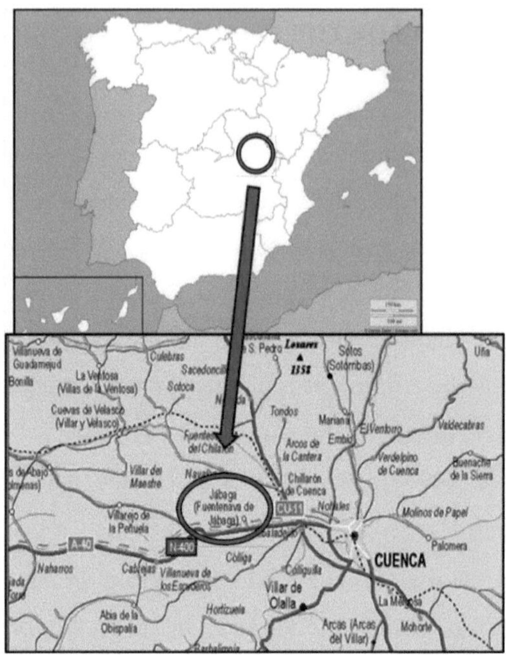

Figure 1. Geographical location of the sampled pine forest in the Iberian Peninsula.

The pine forest studied is located on Oligocene limestone sedimentary deposits, originating from the drying out of lake basins formed by marl, gypsum and limestone. The soil that supports the pine forest is a Terra Rossa, also defined as a brown limestone soil on poorly consolidated material, with areas stony and a poorly developed humus horizon (RIVAS-MARTÍNEZ, 1987), on a stony surface with an uneven topology (ORTIZ VALBUENA, 1992).

Figure 2. Geographical location of the study area. The delimited area of the Fuentenava de Jábaga can be seen, where the pine forest sampled is in shaded colour.

VEGETATION

The vegetation of the study area belongs to the supra-mesomediterranean floor and to the Castilian-Alcarreño-Manchegan basophilic biogeographic region of Quercus faginea or quejigo, whose potential vegetation is the gall-oak groves. The predominant species are Pinus nigra Arn. and gall oak (Quercus faginea Lam.) RIVAS- MARTÍNEZ, 1987).

The black pine (Pinus nigra) is an orophilous sub-Mediterranean species that is highly resistant to low temperatures and severe summer droughts. For this reason, it is a species with a cold-steppe tendency, although due to lithological and geomorphological compensations it is capable of living under a wider range of climatic conditions. Pinus nigra normally occupies rocky outcrops, ridges and steep slopes where soils are not very developed. It has the ability to colonise unfavourable terrain and plays a role important in soil formation (REGATO & ESCUDERO, 1989). The black pine stands in the Serranía de Cuenca are among the largest and best preserved in the Iberian Peninsula (MONERO et al., 2010).

Species such as Acer monpessulanum L., Sorbus aria (L.) Crantz and Quercus humilis Mill. can also be observed in the area. In the shrub stratum we find Amelanchier ovalis Medik., Prunus mahaleb L., Buxus sempervirens L. and Rhamnus saxatilis (Host.) Sibth & Sm. Herbaceous species include the orchids Cephalanthera rubra (L.) Rich., C. damasonium (Mill.) Druce, C. longifolia (L.) Fritsch and Orchis morio L., as well as specimens of Lathyrus filiformis (Lam.) Gay, Hepatica nobilis L., Tanacetum corymbosum (L.) Sch. Bip., Geranium sanguineum L. and Helleborus foetidus L. (CERRO & LUCAS, 2007).

In the pine forest that forms the sampling area, a fire occurred on 21 August 1991, affecting 185 hectares of forest. At the time of sampling (December 2002), some trees still had burnt bark, but the upper part of the canopy was intact. There was no leaf litter in the affected area and the soil lacked horizon differentiation. The vegetation consisted mainly of Pinus nigra and Quercus ilex, which are slow to regenerate. There were also specimens of Thymus vulgaris L., Satureja intricada Lange in Vidensk, Salvia lavandulifolia Vahl, Aphyllantes monspeliensis L. and Fumana ericoides (Cav.) Gand. in Magnier (CERRO & LUCAS, 2007).

CLIMATOLOGY

The climate of the Serranía is characterised by its Mediterranean type, moderated by the altitude and the orographic effect of its relief, which exposes it to humid westerly winds. It also has a certain continental character. Rainfall peaks in autumn, with November being the wettest month. The temperatures are more cold in winter while that the summer is not . excessively warm, with July being the hottest month (DOMÍNGUEZ-FONTANA, 2011). The average annual temperature is 11.5°C, and rainfall is 583 mm, with an average summer rainfall of 96 mm (RIVAS-MARTÍNEZ, 1987).

4. MATERIAL AND METHODS

4.1 SAMPLE COLLECTION

Sampling was carried out between 14 and 15 December 2002, so it corresponds to autumn sampling (O). Thirty samples were studied, of which 11 belong to natural soil (N), 14 correspond to samples of burnt soil (I), and 5 to the edge zone (B). Three plots separated by at least 50 m were selected: one in natural forest (N), one in burnt forest (I) and one in a transition or edge zone (B). In each plot, 6 samples of the different soil types (natural, burned and edge) were collected. For this purpose, trees or stumps remaining in the area after the fire were located and the extraction area was established at a distance of 2 metres from the foot. At each point a litter sample (H) was taken, when present, one at surface (S) and one at depth (P). When moss was found in the natural area, it was removed sampling. The samples were collected on a surface of 10x10 cm and contain a volume of 500 cc of material. They were then numbered from 1 to 6 and the ambient and soil temperatures were measured using a thermometer.

4.2 EXTRACTION, PREPARATION AND IDENTIFICATION OF SPECIMENS

The extraction of the specimens was carried out using the Berlese-Tüllgren system. The soil sample was placed in a sieve over a funnel, above which a permanent light was kept shining on the sample. for 15 days. The edaphic fauna tends to flee from light and dryness, as they live in a dark and humid environment, so they pass through the sieve and go down the funnel until they fall into a container with 70% alcohol, which acts as a preservative (BARRIENTOS, 2004) **(figure 3).** The fauna was then separated from the sample with the aid of a paintbrush and an Olympus binocular magnifying glass at 150X. The specimens separated by orders and placed in tubes with 70% alcohol were labelled in order to preserve them for future taxonomic studies.

For the preparation of springtails, it is necessary to previously immerse them in lactic acid for at least one week. This compound softens the tissues and musculature and at the same time clears the cuticle of the animal, facilitating the observation of the structures necessary for identification at species level (PALACIOS-VARGAS & MEJÍA, 2007).

Figure 3. Apparatus with funnels of the Berlese-Tullgren system for the extraction of soil fauna.

The specimens were then mounted in semi-permanent preparations using Höyer mounting fluid (distilled H_2O, chloral hydrate, glycerine and gum arabic). For identification, a phase contrast microscope was used at 400 and 1000 magnification, because many of the basic characters and structures are not visible at lower magnifications. Handbooks and dichotomous keys are used for taxonomic identification (BARRIENTOS, 2004; DINDAL, 1990; JORDANA & ARBEA, 1989; JORDANA et al., 1997; POTAPOW, 2001), as well as specialised literature.

4.3 DATA ANALYSIS

4.3.1 DIVERSITY INDICES

In order to carry out a comparative study of the springtail communities present in the different soil levels of all sampled sites in both natural and burned forest, different diversity indices appropriate for specific studies have been used.

Shannon-Wiener Index ($H = -\Sigma pi - \ln pi$). It is a measure used to assess specific biodiversity, as it takes into account both the number of species and their relative abundance in a community. It provides information on the heterogeneity of the community and the degree of uncertainty associated with random selection of individuals (PLA, 2006).Shannon-Wiener index values vary between zero, when only one species is present, and the logarithm of the number of species, when all groups are represented by the same number of individuals (MAGURRAN, 1988; MORENO 2000). When the index is close to zero, it indicates that a community has low diversity and one species predominates over the others, whereas, as the index approaches the logarithm of the number of species, the community is more diverse and all species are equally present.

Specific richness ($Hmax = \ln S$). Provides a measure based solely on the number of species present in a sample. The higher the value of Hmax, the higher the specific richness. Maximum diversity is achieved when all species are equally represented. However, this index does not take into account the relative importance of each species (PLA, 2006).

Pielou's equity index ($J = H/Hmax$). This is a measure that indicates the ratio of observed diversity to the maximum expected diversity. The index ranges in value from 0 to 1. A value of 1 indicates that all species present in the sample are equally abundant, which means that all species present in the sample are equally abundant. implies a perfectly equal distribution. This scenario is achieved when all species have the same number of individuals (MAGURRAN, 1988; MORENO 2000).

Simpson's dominance index ($\lambda = \Sigma pi^2$). A measure that provides information on the probability that two randomly selected individuals are of the same species. It is inversely proportional to fairness and is influenced by the importance of dominant species (MAGURRAN, 1988; MORENO 2000). A value of 1 indicates low diversity and high dominance, while as the index increases, diversity decreases.

4.3.2 MULTIVARIATE ANALYSIS

With all the data obtained, a series of multivariate statistical analyses were carried out using the SPSS 28.0 statistical package. The three types of analysis were carried out with the different groups of edaphic fauna separated, and then with the different species of springtails.

Spearman Correlation Index. The coefficient measures the degree of relationship or association that usually exists between two random variables (RESTREPO & GONZÁLEZ, 2007). It is used when we are dealing with the analysis of variables that have a discrete quantitative nature and/or are hierarchical (SALINAS, 2007).

Principal Component Analysis. This is a statistical-algebraic technique for reducing the number of variables, which seeks to summarise and organise the information present in a data matrix. The process involves transforming the data matrix into a vector space, where axes or dimensions are sought that are a linear combination of the variables introduced (COLINA & ROLDÁN, 1991).

Discriminant Analysis. It is used to correctly classify subjects or objects and helps to identify the most relevant variables for an accurate classification (TORRADO & BERLANGA, 2013). The variables that have been taken as grouping variables are: the state of the ecosystem (natural/burnt/edge), and the season (autumn/winter) only for the study of edaphic fauna.

5. RESULTS OF THE STUDY OF SOIL FAUNA

5.1 FAUNISTIC SURVEY

A total of 11812 individuals belonging to 21 different taxonomic groups were found, the distribution of which, according to the area from which the sample was taken, can be seen in tables **1, 2 and 3 in the appendix.**

This section analyses the most representative biological and ecological characteristics of the classified groups.

PHYLUM ARTHROPODA

Subphylum Chelicerata

Order Acari. It is one of the most representative and abundant groups of edaphic fauna. Their specific diversity is high due to the great variability of life forms they present and the different environmental conditions in which they can survive. This is why they can be used as bioindicators of ecosystem disturbance (SOCARRAS, 2013).

The suborder **Oribatidae** (Cryptostigmata) is one of the most common groups found in organic horizons in most regions of the world. These mites are most abundant in forest soils little disturbed by human activity, generally with a low pH (PETERSEN, 2002; ARROYO et al., 2003). They contribute to the decomposition of organic matter by fragmenting it and facilitating the action of microorganisms (SOCARRAS, 2013). Some species migrate towards the leaf litter or fermentation level in winter and are distributed in groups depending on the amount of nutrients available (DINDAL, 1990).

Order Pseudoscorpionida. They are predatory animals that regulate the populations of microfauna and mesofauna in the soil, feeding on mites and springtails among others (PARISI, 1979). They live in diverse environments, generally humicolous, such as under tree bark, in fungi, on mosses and leaf litter, in soil crevices and rocks (BARRIENTOS, 2004; VILLEGAS-GUZMÁN & PÉREZ, 2005).

Subphylum Hexapoda

Order Coleoptera. It is the order with the most species in the animal kingdom, and with a great morphological diversity (CROWSON, 1981). It includes aquatic and terrestrial species. The latter usually inhabit trees, shrubs, grasses, mosses or lichens. Their food sources are very diverse: plants, decomposing organic matter or other species. Some are pests, with larvae causing the most damage to crops, and others establish ectosymbiosis relationships with fungi, mites and nematodes (ALONSO-ZARAZAGA, 2015). Both adult individuals and larval specimens have been found.

Order Collembola. They are the most numerous group in the edaphofauna, together with the mites. They feed on plant debris and decomposing organic matter (RICHARDS & DAVIES, 1984), so they play a crucial role in nutrient recycling and favour bacterial and fungal decomposition (LUCIÁÑEZ & INIESTO, 2006). They can also feed on pathogenic fungi, reducing their populations and favouring plant growth (SOCARRAS, 2013). They are usually found among the leaf litter on the ground or under the bark of trees, and their population increases with cold temperatures (DINDAL, 1990). Consequently, they have developed various strategies, including ecomorphosis and cyclomorphosis, to adapt to extremes of temperature and humidity, as well as seasonal variations (CUTZ-POOL et al., 2003; ARBEA & BLASCO-ZUMETA, 2001). As they require very specific conditions of temperature, humidity and pH, can be used as indicators of soil environmental quality and can reveal information the evolution of ecosystems under different degrees of disturbance (SOCARRAS, 2013).

Order Dermaptera. They are omnivorous animals, although some species can be predatory. They can be found throughout the year depending on the climate, although they are most prevalent in June and September (HERRERA, 2015).

Order Diptera. They are cosmopolitan animals that can be found in both terrestrial and freshwater habitats, being one of the most ecologically diverse groups (HJORTH-ANDERSEN, 2015). The diet of larvae and adults, as well as the environment in which they live, is highly variable. As in order Coleoptera, both adults and larvae are present in the samples studied.

Order Embioptera. Their modified anterior tarsi equipped with silk-producing glands allow them to create silk galleries that serve as shelters for females that remain alongside eggs and nymphs. Some studies suggest that adult males do not feed, while nymphs and females feed on leaf litter, tree bark, lichens or mosses (TORRALBA-BURRIAL, 2015).

Order Hemiptera. They are the most extensive order of pachymetamorphic insects. Most are phytophagous, but there may be predatory species (CHANDRA, 2008). They inhabit leaf litter, stones or cracks in the substrate and contribute to the fragmentation of the organic layer by digging tunnels, fragmenting leaves or sucking roots (ALVARADO & SELGA, 1961; GOULA & MATA, 2015).

Order Hymenoptera. A highly diversified order with both phytophagous and predatory or parasitic species (PUJADE-VILLAR & FERNÁNDEZ GUYABO, 2004). Most of the specimens found belonged to the family Formicidae, the ants, which constitute a very important link in the soil ecosystem, many species being able to select certain conditions of temperature and humidity (RUIZ, 1999), and can also be used as bio-indicators of soil quality given their importance in promoting essential soil processes for its formation and preventing its degradation (DORAN et al., 1994).

Order Protura. Cosmopolitan and exclusively edaphic animals. They live in areas rich in organic matter and with high humidity, in the deepest strata, so they are not usually exposed to the alteration of the upper strata (DINDAL, 1990; BARRIENTOS, 2004; SOCARRÁS, 2013). Some of these animals may feed on mycorrhizae.

Order Psocoptera. They inhabit leaf litter, tree trunks, soil organic matter, fungi and lichens, which basically constitute their diet (RICHARDS & DAVIES, 1984). They are abundant in drought conditions and are pioneers in the recolonisation of disturbed or disturbed areas, indicating a progressive recovery of the soil (SOCARRAS, 2013).

Order Thysanoptera. They inhabit vegetation, tree bark or decaying plant debris. Predatory species regulate natural populations of both Thysanoptera and other orders (MCGAVIN, 2001), although species can also be found feeding on leaves, pollen or sap from within living plant tissues (RICHARDS & DAVIES, 1984). Some species act as pollinators, moving within the plant itself, between adjacent plants or being transported by the wind to another plant (GOLDARAZENA, 2015).

Subphylum Myriapoda

Class Chilopoda. These are predatory animals that reside on the ground in dark places, such as under leaf litter, fallen logs or rocks. In the samples analysed, only individuals of the order **Geophilomorpha** were found. burrowers and mostly carnivores, although their diet may also include vegetation (VOIGTLANDER, 2001).

Class Pauropoda. They are rare animals, as they do not tolerate large environmental variations and are very sensitive to agricultural practices (SOCARRAS, 2013). They can be found in tree trunks, leaf litter, mosses or under stones, in places with stable temperatures and humidity (SCHELLER, 1990). They feed on organic debris (**figure 4**).

Figure 4. Morphotype of a Pauuropod (left image). Morphotype of a Synphyle (right). Note the euhedral morphology of both groups characteristic of deep soils (Pauuropod: taken from inaturalist.org. Even Dankowicz (some rights reserved) (Symphylum: taken from https://encyclopediaofarkansas.net/)

Class Symphyla. They feed on decomposing organic matter, so they play a very important role in the soil ecosystem, although some are phytophagous and can be considered pests for crops (DINDAL, 1990). These animals have a fragile body, so they sometimes use galleries and tunnels built by other animals, and live in humus, soil, leaf litter, moss or decaying wood.

Class Diplopoda. Specimens of two orders were identified in this study: **Julida** and **Polyxenida**. Julida are herbivores and are commonly found in areas with high humidity, under rocks, trunks and leaves, where they contribute to the fragmentation of plant remains (BARRIENTOS, 2004). Polyxenids are also phytophagous, but are more common in dry environments, under tree bark and stones (WRIGHT & WESTH, 2006).

PHILO ANNELIDA

Oligochaeta subclass. These animals play a fundamental role in the soil ecosystem as they improve soil fertility by modifying the physical, chemical and biological characteristics, changing the texture and participating in the decomposition of organic matter and the regulation of biogeochemical cycles (EGERT et al., 2004). Thanks to their activity, they aerate the substrate and favour the entry of water.

FILO NEMATODA

These organisms are microflora consumers or saprophytes, which influence the decomposition and release of nutrients (NAVIA et al., 2006). It is the most abundant phylum of the edaphic fauna (ZANCADA & SÁNCHEZ, 1994), however, only 3 specimens were found because the Berlese funnel **(figure 3)** mainly collects specimens that live on oxygen from the soil air and not from water, like this group of animals.

5.2 NUMERICAL ANALYSIS OF SOIL FAUNA GROUPS

The 11812 specimens corresponding to the 30 autumn samples were analysed. Of these, 7497 individuals correspond to 17 different taxonomic groups collected from natural soil, constituting 63.47% of the total. In the burned soil, 2375 specimens from 14 taxonomic groups were obtained, which is equivalent to 20.11% of the total, and in the edge or transition zone, 1940 specimens belonging to 12 taxonomic groups were studied, which represents 16.42%. Mites are the most numerous group, representing 80.63% of the total, and springtails represent 17.08%, so together they account for 97.71% of all the specimens found. **Figure 5** shows the counts of mites, springtails and other animals according to the different types of forest (natural, fire or edge).

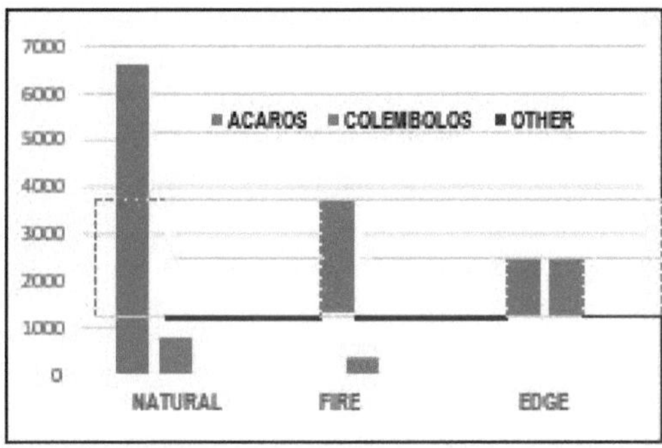

Figure 5. Representation of the number of mites, springtails and other taxonomic groups according to forest type (natural, fire or edge).

It can be seen that while in the natural forest mites dominate notably, in the transition zone the mite and springtail communities are almost equal in abundance.are practically equal in abundance. The rest of the taxonomic groups are mainly observed in the natural and burnt soil samples, being slightly higher in the burnt area.

Figure 6 shows the count of specimens for each taxonomic group, excluding mites and springtails. Pauropods stand out as the most abundant order, representing 27.61% of the total, being more abundant in the burnt soil. The next most numerous groups are the larvae of both Diptera (19.03%) and Coleoptera (9.33%) and symphylans (12.69%).

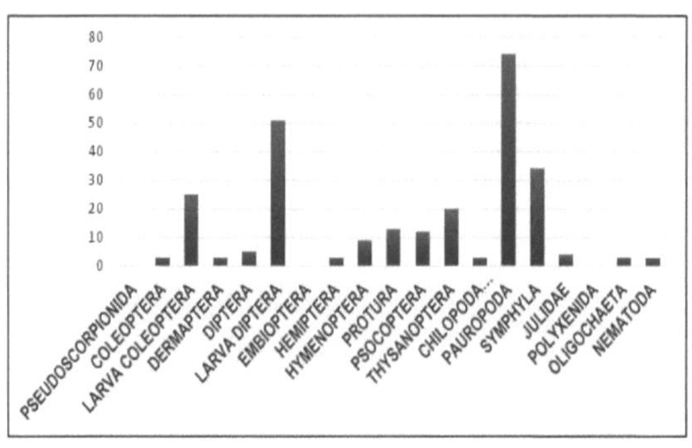

Figure 6. Total number of specimens by faunal groups found.

5.3 STATISTICAL ANALYSIS OF SOIL FAUNA GROUPS

5.3.1 SPEARMAN CORRELATION INDEX

A list of the significant correlations obtained after Spearman's correlation analysis between the different faunal groups sampled in autumn can be found **in table 4 of the appendix**. Although all significant correlations are represented, it is worth noting the value between pseudoscorpions and polyxenids (1), between springtails and psocoptera (0.51), and the negative correlation between mites and chylopods (-0.388). Temperature does not correlate significantly with any group of edaphofauna.

5.3.2 PRINCIPAL COMPONENT ANALYSIS

Figure 7 shows the graphical representation of the principal component analysis for the soil fauna. The first component, represented on the X-axis, absorbs 15.99% of the variance, and the second, on the Y-axis, 11.37%. The total cumulative extraction of the first 4 components was 46.72%. The graph shows the distribution of faunal groups according to their presence and abundance in the natural forest, the edge and the burnt area. It can be seen that in the positive region of the X axis are the most abundant orders in the burnt forest, such as the symphylans and pauropods. Organisms from the natural forest, such as springtails or mites, are distributed especially in the negative region of the two axes. The most abundant individuals in the edge or transition zone, such as psocoptera or chylopods, are close to the axis in the negative region of the X-axis.

5.3.3 DISCRIMINANT ANALYSIS

Discriminant analysis uses soil condition (natural, edge or fire) as a grouping variable and selects several groups of discriminant fauna: springtails, pauropods, mites and oligochaetes. The first three are obtained as discriminating because of their greater abundance in the burnt soil in relation to the other faunal groups. Springtails and mites are dominant, with some exceptions, in any given soil. Oligochaetes are selected as a characteristic group of the transition zone.

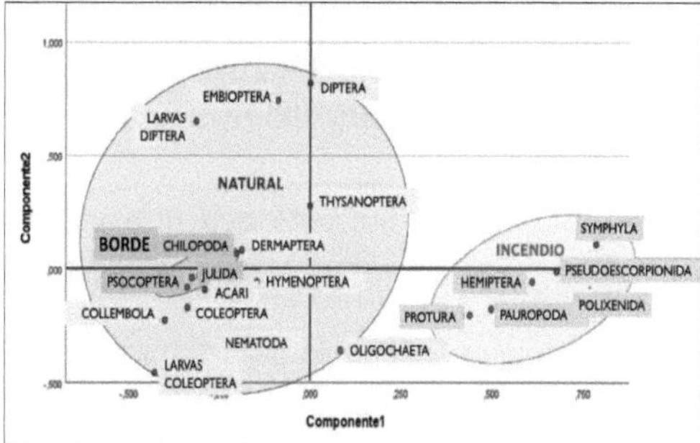

Figure 7. Graphical representation of the principal component analysis of the soil fauna of the autumn sampling for the first two components.

STUDY OF THE FAUNA OF SPRINGTAILS

5.4. STUDY OF SPRINGTAIL COMMUNITIES

A total of 2110 springtails belonging to 11 families and 33 species were collected. Of these, 833 specimens of 20 species belong to the natural forest, 350 of 20 species to the burnt forest, and 927 specimens of 19 species to the transition zone between the two. **Tables 5, 6 and 7 in the annex** show the abundance of the species in each of the samples studied.

5.4.1 FAUNAL SURVEY

Due to the great adaptation of springtail species to the habitat in which they live, it is useful to give a brief systematic synopsis of the species encountered, commenting briefly on their ecology and distribution.

FAMILY HYPOGASTRURIDAE (BÖRNER, 1906)

Ceratophysella tergilobata (Cassagnau, 1954)

An infrequent hemiedaphic species but collected in a wide range of different forests, from holm oak and juniper to eucalyptus (LUCIÁÑEZ & INIESTO, 2006). It is found as a dominant species in pine forests of P. pinaster Aiton, in the Sierra de Gredos. It adapts easily to extreme dry conditions by developing protective cuticular structures (JORDANA et al., 1997). It is also considered a Mediterranean trogloxena (ARBEA et al., 2021). Found in Europe and part of East Asia. In Spain it has so far been reported from Tamajón (SIMÓN, 1985), Fuentelahiguera (LUCIÁÑEZ & SIMÓN, 1987), Navarra (ARDANAZ & JORDANA, 1983), Arenas de San Pedro in Gredos (LUCIÁÑEZ & INIESTO, 2006) and Andalusia.

Hypogastrura purpurescens (Lubbock, 1868)

Species with a wide distribution and ubiquitous ecology. It has been found under bark, trunks, grasslands, cultivated soils, pine forests, oak forests, and in litter. It has been found in cave environments (ARBEA et al., 2021). It has a cosmopolitan distribution.

Microgastrura duodecimoculata Stach, 1922

It is a hemiedaphic species although it can go deep into the substratum and is also adapted to the cave environment. It is found in forests of various trees such as cedars, pines or cork oaks, as well as in soils with humus and leaf litter (LUCIÁÑEZ & SIMÓN, 1989). It is distributed throughout Europe.

Xenylla schillei Börner, 1903

Hemiedaphic species that tends to be found in the superficial layers of the soil, as well as in mosses in meadows and deciduous forests (JORDANA et al., 1990). It is typical of open environments. The species has been found in central and southern Europe. It is widely distributed on the Iberian Peninsula (JORDANA et al., 1997).

Xenyllogastrura octoculata (Steiner, 1955)

Abundant in soils, mosses and lichens both in forests, especially pine forests, and in meadows. It is an euedaphic species associated with calcareous soils and a Mediterranean environment (JORDANA et al., 1990). It can be found on the ground, mosses or lichens in forests and meadows. A species with a south-European and Mediterranean distribution. In the Iberian Peninsula it is found in the northern half, its type locality being Aranjuez.

FAMILY BRACHYSTOMELLIDAE STACH, 1949

Brachystomella parvula (Schäffer, 1896)

It lives in humid areas and open spaces, young pine forests and cleared forests (LUCIAÑEZ, 1990). It is a species that resists dryness well (GAMA et al.,

1989) and can survive for long periods of time in a state of anhydrobiosis (ARBEA & BLASCO-ZUMETA, 2001). It can be found in caves accidentally (ARBEA et al., 2021).It is a cosmopolitan species.

FAMILY NEANURIDAE (BÖRNER, 1901)

Friesea steineri Simon, 1975

It has been found in the leaf litter of oak groves, in savin, thyme and grass meadows, holm oak and rockrose litter in dehesas.

Species endemic to the Iberian Peninsula, in its central and southern region (JORDANA et al., 1997).

Bilobella aurantiaca (Caroli, 1912)

A species found in various types of forest, preferring areas that are not very humid and able to withstand periods of drought. It can be found in mosses, leaf litter and soil of forest stands that include pines, holm oaks or cork oaks among others (DALLAI, 1973; JORDANA et al., 1997). It has a Palaearctic and Mediterranean distribution.

Micranurida pygmaea Börner, 1901

Holarctic species found in low mountain localities associated with pine forests. RUSEK (1998) and JORDANA et al. (1997) cite it from acid forest environments. It is mainly distributed in the north and centre of the Iberian Peninsula, although it has also been found in the Balearic Islands.

Pseudachorudina sp.

It has not been possible to arrive at a species level as it is a young specimen that has not yet developed the structures necessary for its identification.

Pseudachorutes parvulus Börner, 1901

It inhabits the surface layers of leaf litter (POTAPOW et al., 2016), mosses, lichens and under bark. It has the ability to survive in a state of anhydrobiosis (ARBEA & BLASCO-ZUMETA, 2001) so it adapts to the extreme conditions of biotopes. Its distribution is cosmopolitan.

Simonachorutes romeroi (Simon, 1986)

It has been found associated with the leaf litter of Juniperus thurifera forests, and thickets of Genista scorpius, Echinospartum horridum and Buxus sempervirens. It is a species of Iberian distribution.

FAMILY ODONTELLIDAE MASSOUD, 1967

Odontellina nivalis (Cassagnau, 1959)

It is a euedophagous animal, although the morphology of its styliform mouthparts allows for another type of suction feeding (ARBEA, 1988). It has a Mediterranean, European distribution.

Superodontella selgae (Arbea, 1990)

It has been found in leaf litter and heath humus, ferns and larch woodland. It is an Iberian species, distributed in the northern half of Spain.

FAMILY ONYCHIURIDAE BÖRNER, 1901

Protaphorura armata (Tullberg, 1869)

Euedaphic species, found both in meadows and forest stands. Possibly cosmopolitan in distribution.

FAMILY TULLBERGIIDAE BAGNALL, 1935

Mesaphorura macrochaeta Rusek, 1976

It is a euhedral and cosmopolitan species (**figure 8**). It inhabits acid soils with little biological activity, dry meadows, coniferous and oak forests or alpine areas (LUCIÁÑEZ & INIESTO, 2006). It is considered troglophilic (ARBEA et al., 2021). It is possibly cosmopolitan in distribution. In the Iberian Peninsula it can be found in almost any region.

Wankeliella medialis Simón & Jordana, 1994

Euedaphic species, inhabiting deep podzolic soils in beech forest clearings. This is a species endemic to the Iberian Peninsula, and this is the second record after the description in the Sierra de Urbasa in Navarre.

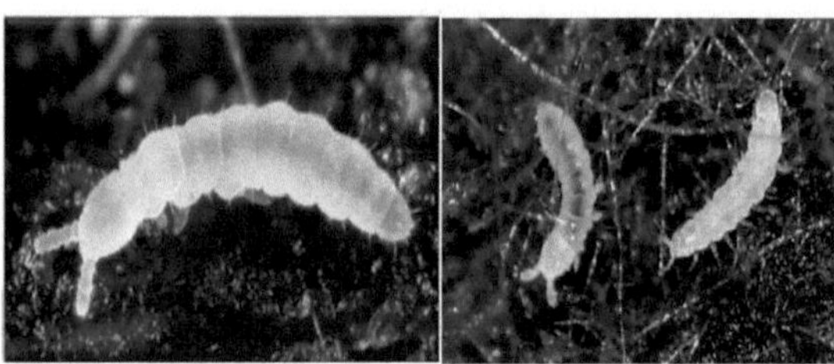

Figure 8. Morphotype of Mesaphorura macrochaeta (left image). On the right, individual of the family Onychiuridae next to a Pauuropod (taken from bugguide.net. Copyright212 Josh D. Kouri).

FAMILY ENTOMOBRYIDAE SCHAËFER, 1896

Entomobrya multifasciata (Tullberg, 1871)

A mesophilic, hemi-daphic and atmobic species. It is a rare species found in native vegetation, and is therefore frequently found in pastures and disturbed soils (FJELLBERG, 2007). The species is cosmopolitan in distribution.

Entomobrya nicoleti (Lubbock, 1868)

Lives in leaf litter and on the forest floor. Palaearctic in distribution, although possibly more widespread (FJELLBERG, 2007).

Lepidocyrtus lusitanicus Gama, 1964

It usually occurs in forests of cedar, eucalyptus, acacia, pine, pyrethrum and on moss (LUCIÁÑEZ & SIMÓN, 1989). Specimens live on herbaceous vegetation or on the ground. They can inhabit various types of herbaceous vegetation of forest or beach, and a very wide altitudinal range (from the level of the sea up to 1800 metres above sea level) MATEOS, 2008). Its distribution is restricted to the Iberian Peninsula.

Pseudosinella sp.

The genus is euhedral. The specimen found could be a new species (it has 4+4 eyes). It is juvenile and therefore certain structures necessary for identification have not developed.

FAMILY ORCHESELLIDAE BÖRNER, 1906

Heteromurus major (Moniez, 1889)

They have a troglophilic tendency and can therefore be found under stones or leaf litter, and tend to inhabit forests of different species. It seems to be found more frequently in forests of fagaceae (LUCIÁÑEZ & SIMÓN, 1989; LUCIÁÑEZ & INIESTO, 2006). It is distributed throughout Europe and the Mediterranean region.

FAMILY ISOTOMIDAE (BÖRNER, 1913)

Ballistura navacerradensis (Selga, 1962)

It is a psammophilous species. Found in grasslands, coastal biotopes and dunes.

European distribution.

Folsomides parvulus Stach, 1922

Xerophilic, although it has also been found in forests and meadows. It can survive in a state of anhydrobiosis (ARBEA & BLASCO-ZUMETA, 2001).

Cosmopolitan distribution (BABENKO et al., 2019).

Hemisotoma thermophila (Axelson, 1900)

Species tolerant to environmental variation, xerophilic, nitrophilic and thermophilic (ARBEA & JORDANA, 1990; ARBEA & BLASCO-ZUMETA, 2001). Preferably inhabits the surface layers, but can also be seen in the first centimetres of the soil. mineral soil (ARBEA & JORDANA, 1990). According to FJELLBERG (2007) it is abundant in environments with a high organic matter content (**Figure 9**). Cosmopolitan species.

Isotoma viridis Bourlet, 1839

Found on leaf litter and forest floor, on arable land. Abundant in oak groves. Holarctic species (BABENKO et al., 2019).

Isotomiella minor (Schäffer, 1896)

Euhedral species (POTAPOW et al., 2016) and sensitive to pollutants (POTAPOW, 2001). It is acidophobic although it can inhabit very diverse biotopes (LUCIÁÑEZ & SIMÓN, 1989; RUSEK, 1998). It is also associated with the cave environment. It is a cosmopolitan species.

Isotomodes bisetosus Cassagnau, 1959

Holarctic and euhedral species typical of disturbed areas (FJELLBERG, 2007). It has been found in the Sierra de Guadarrama (LUCIÁÑEZ & SIMÓN, 1991) and Navarra (JORDANA et al., 1997) among others.

Isotomurus sp.

Epigeous, hygrophilous or aquatic organisms, which can feed on algae and can survive desiccation (RUSEK, 1998). This genus is commonly found in wet meadows, mountainous areas, lakes and rivers. However, it has also been recorded in pine, juniper and oak forests (LUCIAÑEZ, 1990). The specimens of the species found are juvenile forms so it has not been possible to identify the species.

Parisotoma notabilis (Schäffer, 1896)

Holarctic and hemiarctic species that can live in both natural conditions and disturbed areas (POTAPOW et al., 2016). It has a cosmopolitan distribution.

Tetracanthella pilosa Schött, 1891

Lives in mosses on rock and tree trunks in lowland and low mountainous areas. Holarctic, European species according to POTAPOW (2001) (**figure 9**).

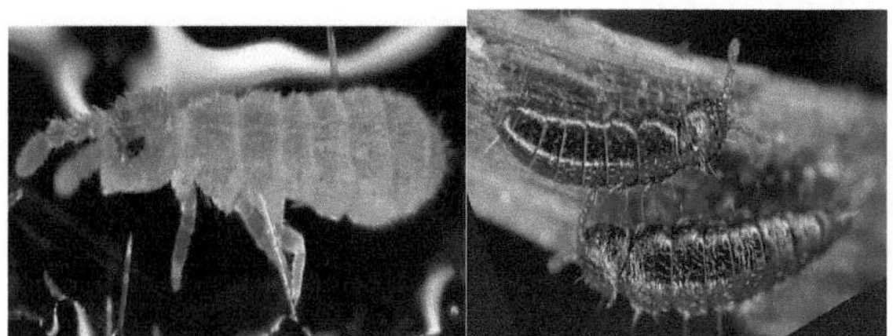

Figure 9. Hemisotoma thermophila (Source: collembola.org Garcelon, P.). Tetracanthella pilosa (Source: collembola.org Huskens, M.L.)

FAMILY NEELIDAE FOLSOM, 1896

Megalothorax minimus Willem, 1900

A cosmopolitan, euhedral and troglophic species, adapted to living in the deepest strata of the soil. It is typical of forest soils and has been found in the leaf litter of conifers, mosses of beech, oak and pine forests (BONNET et al., 1979).

FAMILY SMINTHURIDIDAE BOERNER, 1906

Sphaeridia pumilis (Krausbauer, 1898)

A hemideciduous and mesophilic species, inhabiting both leaf litter and soil. It lives in humid environments, leaf litter and forest floor, such as rainforests, or in aquatic environments, and survives periods of drought in the egg stage (ARBEA & BLASCO-ZUMETA, 2001; FJELLBERG 2007). Holarctic distribution.

5.4.2 NUMERICAL STUDY

The most abundant species is Tetracanthella pilosa, which constitutes 34.80% of the total. It is followed in abundance by Mesaphorura macrochaeta (20.35%) and Ceratophysella tergilobata (11.38%). These three species represent 66.53% of the individuals collected.

Table 1 shows the most abundant species (in terms of relative abundance) in each of the ecosystems studied.

Table 1. Relative abundance of the most abundant species by forest type.

Natural		Fire		Border	
Species	Abundance	Species	Abundance	Species	Abundance
M. macrochaeta	30,01%	M. macrochaeta	37,71%	T. pilosa	73,62%
C. tergilobata	26,17%	W. medialis	12,00%	H. thermophila	7,35%
P. notabilis	21,49%	P. armata	10,86%	M. macrochaeta	6,92%

The percentage of each of the biogeographical categories for the species found is presented below (**figure 10**). There is a clear predominance of widely distributed, cosmopolitan, Holarctic and Palaearctic species, but the 19% of endemic or Iberian species is not negligible.

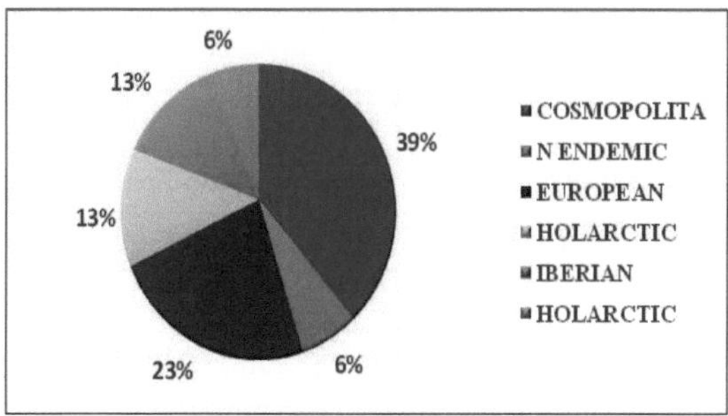

Figure 10. Biogeographical distribution of springtail species according to their percentage of occurrence.

5.5. DIVERSITY INDICES

A study was carried out of springtail diversity indices according to the type of soil sample (natural, fire or edge). The results obtained are shown in **figure 11.** It can be seen that diversity, as measured by the Shannon-Wienner index, is higher in the burnt forest. Specific richness shows very similar values in the three ecosystems. Evenness or uniformity is higher in the burned soil, and the values of Simpson's index, which represents dominance, are higher in the transition zone due to the abundant population of T. pilosa.

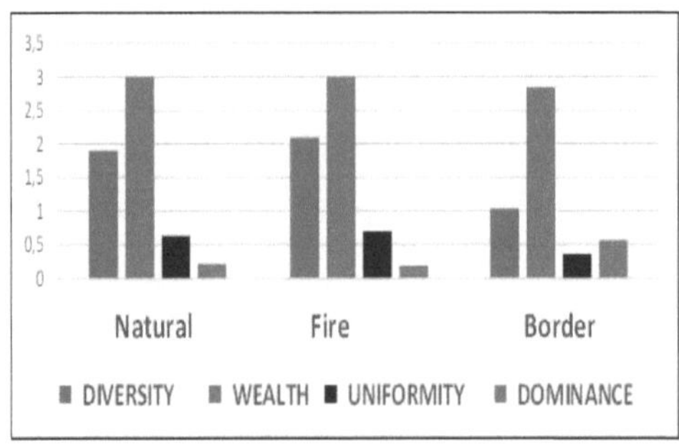

Figure 11. Diversity index values according to sample type (natural, fire or edge).

5.6 STATISTICAL ANALYSIS OF SPRINGTAIL POPULATIONS

5.6.1 SPEARMAN CORRELATION INDEX

The significant correlations obtained after calculating Spearman's correlation index between the different species of springtails found in autumn can be seen in **table 8 in the appendix**. The value of 1 between Hypogastrura purpurescens and Entomobrya nicoleti, and between Xenylla schillei and Superodontella selgae stands out. The reason is their common presence and abundance in the same samples. Isotomodes bisetosus with Sphaeridia pumilis, and with P. notabilis are the two pairs with negative correlation.

5.6.2 PRINCIPAL COMPONENT ANALYSIS

Figure 12 shows the graphical representation of the principal component analysis for the springtail species collected in the autumn sampling. Component 1 is represented on the y-axis (17.2% variance) and the y-axis representing component 2 absorbs 12.01%. The total cumulative extraction of the first 4 components was 48.85%.

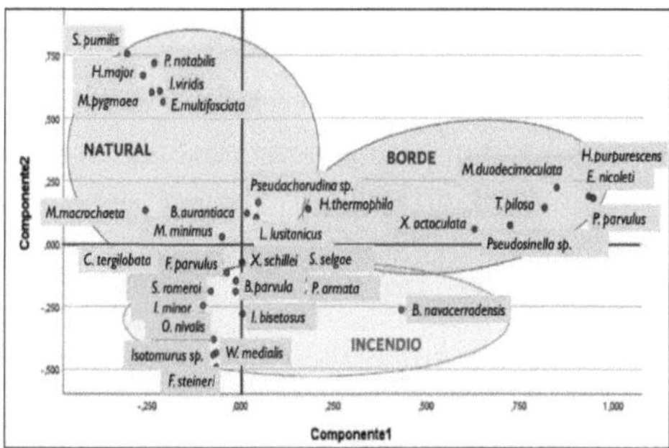

Figure 12. Graphical representation of the principal component analysis of the springtails from the autumn sampling for the first two components.

The distribution of springtail species can be observed in three groupings corresponding to the three areas sampled, the natural forest, the transition or edge zone and the burned area. The species present or more abundant in the natural zone, such as Mesaphorura macrochaeta or Isotoma viridis, are located in the negative region of the X-axis and the positive region of the Y-axis. The most abundant species in the transition zone, such as Tetrachantella pilosa or Hemisotoma thermophila, are represented in the positive region of the X-axis. The characteristic species of the burned area, such as Pseudachorutes romeroi or Isotomodes bisetosus, are predominantly concentrated in the negative region of the Y-axis.

5.6.3 DISCRIMINANT ANALYSIS

For this analysis, the grouping variable was soil condition (natural, edge or burnt). Pseudachorutes parvulus, Hypogastrura purpurescens, Hemisotoma thermophila and Tetrachantella pilosa, all characteristic of the transition zone, were selected as discriminating variables. Isotomurus sp. is the species selected for the burned ground, and Heteromurus major and Sphaedia pumilis are characteristic of the natural forest.

6. DISCUSSION

From the data obtained after the various numerical and statistical analyses it can be seen that fire severely affects soil biota. All taxonomic groups were reduced in number after a standardised fire impact. This confirmed numerous field observations (GONGALSKY et al., 2012) where it can be stated that the type of soil sampled, even within a fire, affects the distribution and abundance of edaphic fauna and collembolan communities. A considerable decrease in the number of individuals present in the burned area compared to the natural forest can be observed. However, the edge zone does not follow the expected pattern as it has a lower number of mites and other edaphic organisms than the burnt soil, but a higher number of springtails than the natural soil. This has already been cited in the literature. The study by SUHADI et al. (2020) shows a greater richness and diversity of springtails in the transition zone between the natural and burnt zones.

In burned soil there is a much lower layer of vegetation and organic matter than in natural forests. This layer plays a fundamental role in providing thermal and water protection to the soil ecosystem and its reduction could explain the reduction in the number of organisms in the burned soil (LUCIÁÑEZ & INIESTO, 2006). The disappearance of leaf litter also causes many organisms to take refuge in the deeper layers of the soil, which leads to a predominance of euhedral organisms. With regard to the edge zone, it is possible that both the mites and some of the edaphic organisms moved to the natural forest zone in search of more stable environmental conditions. On the other hand, springtails may have migrated to the transition zone due to less competition and greater availability of resources than in the natural zone. In addition, some species have certain adaptations that allow them to survive or recover more quickly after a fire. The results of the study confirm that mites and springtails are the most abundant groups in the soil (TEUBEN & SMIDT, 1992). Pauropods are the third most representative faunal group which, together with symphylans, predominate in the burnt soil. They are euedaphic organisms that resist the high temperature conditions produced by fires and colonise the habitats abandoned by other species that do not survive under these conditions. This is why they are considered to be indicators of soil disturbance. In the natural forest, the larvae of both Diptera and Coleoptera stand out, and their total abundance is also very high. Diptera larvae prefer humid environments with a high concentration of organic matter (DINDAL, 1990), and many Coleoptera larvae are phytophagous and can find more plant products in the natural soil. On the other hand, the most representative groups on the edge are the psocoptera and julids, although their abundance is much lower.

Within the study of the springtail fauna, the analysis of the diversity indices reveals that there is greater diversity in the burned soil, and the species richness is the same in both the burned and natural forest. These results could be related to the statements of other authors who have observed a progressive recovery of springtail species in burned soils over time (LUCIÁÑEZ & INIESTO, 2006). On the other hand, uniformity is slightly higher in the burned soil and dominance at the edge. As in the results obtained in the general study of the edaphic fauna, the high correlation obtained with the Spearman stands out, especially that of Isotomodes bisetosus with Sphaeridia pumilis and Parisotoma notabilis. These species can live both in natural conditions and in degraded areas. so they adapt well to almost any type of habitat (FJELLBERG 2007; POTAPOW et al., 2016). It should be noted that both Sphaeridia pumilis and Parisotoma notabilis are hemi-daphic species while Isotomodes bisetosus is a euhedral species, so they will not compete for the same resources. In relation to the abundance and frequency of the species present, as well as discriminant analyses, different sets of species characteristic of the ecosystems studied have been identified. Species such as Heteromurus major or Sphaedia pumilis predominate in the natural soil due to their preference for areas with relatively humid environmental conditions (ARBEA & BLASCO-ZUMETA, 2001). Isotomurus sp. is mostly found in burnt soil as it has a high tolerance to desiccation (RUSEK, 1998). Finally, abundant species such as Tetrachantella pilosa or Hemisotoma thermophila are characteristic edge species due to their high tolerance to environmental variation (ARBEA & JORDANA, 1990; ARBEA & BLASCO-ZUMETA, 2001). In addition, they have dormant eggs with fire-resistant casings that find in these adverse conditions the ideal time to develop and hatch. This has also been seen in another study, still unpublished, carried out with spring samples from the same fire. In that sampling, Xenylla schillei is a very abundant species on the edge that appears almost exclusively in that area due to the hatching of its dormant eggs.

The same is true for some endemic species which, despite having a restricted area of distribution with specific environmental characteristics, are still present after the fire. They remain dormant until conditions are optimal for hatching. This is the case of species such as Friesea steineri or Wankeliella medialis.

From the results of this study, it can be affirmed that the modifications caused by forest fires in the soil organic matter layer, as well as the changes in the soil organic matter layer caused by forest fires, have a significant impact on the soil organic matter layer, as well as on the soil organic matter layer. The specific composition found in the different types of soil samples shows that the communities of edaphic fauna and the abundance and distribution of the communities of springtails, as well as those produced in the vegetation, alter the communities of edaphic fauna and the abundance and distribution of the communities of springtails. Despite this, it is observed that the biological communities themselves are resilient and have diverse strategies to recover their populations.A more complete study would require further sampling in the area in subsequent years to check whether the edaphic communities re-colonise the areas occupied by the burnt soil, and to examine whether the faunal and collembolan composition remains the same or whether there are slight variations. All the data obtained from the autumn, winter, spring and summer surveys could also be pooled to check the differences observed in the distribution and abundance of springtails depending on the season, temperature or humidity.

7. CONCLUSIONS

- In the study carried out in the pine forest of Pinus nigra Arn. in Fuentenava de Jábaga (Cuenca) after the 1991 fire, 11812 individuals belonging to 21 different faunal groups were collected. In the natural pine forest 7497 individuals from 17 different taxonomic groups were found, in the burnt pine forest 2375 specimens from 14 groups, and in the transition zone 1940 specimens belonging to 12 groups.

- Mites are the most abundant group, representing 80.63% of the total, being more numerous in the natural pine forest. Springtails (17.08 %) are more numerous in the transition zone. There is also a significant abundance of pauropods, especially on the burnt ground.

- There is a decrease in the number of individuals present in the burnt area compared to the natural forest. The edge zone, on the other hand, shows a decrease in the number of individuals present in the burnt area compared to the natural forest. less mites and other fauna than burnt soil, and has a higher number of springtails than even natural soil.

- A specific study of the springtail fauna was carried out, collecting a total of 2110 individuals belonging to 33 species. Of these, 833 specimens of 20 species belong to the natural pine forest, 350 individuals of 20 species correspond to the burnt pine forest and 927 individuals of 19 species were found on the edge.

- In the natural forest, M. macrochaeta is the most abundant species (30.01 % of the total), followed by C. tergilobata (26.17 %) and P. notabilis (21.49 %). In the burnt forest the most abundant species are M. macrochaeta (37.71 %), W. medialis (12.00 %) and P. armata (10.86 %) due to its euedaphic character and, in the case of W. medialis, its endemicity. In the border zone, T. pilosa (73.62%), H. thermophila (7.35%), hemiedaphic and Mediterranean, whose dormant eggs hatch because of the high temperatures in this area, and M.

macrochaeta (6.92%), an euhedral and parthenogenetic species, with clear adaptation strategies to soil disturbance.

- Diversity indices were calculated, which indicate higher diversity in burned soil and similar richness in both natural and burned soil. Principal component analyses reveal that the distribution springtail species is influenced by the different edaphic characteristics of the different types of soil samples.

- From the study of abundance and the analyses used, we can see that the distinguishing species in the natural pine forest are Heteromurus major and Sphaeridia pumilis, in the burned area Isotomurus sp and on the edge Pseudachorutes parvulus, Hypogastrura purpurescens, Hemisotoma thermophila and Tetrachantella pilosa, the latter two being the most important.

8. BIBLIOGRAPHY

ALONSO-ZARAZAGA, M.A. (2015). Order Coleoptera. Revista Ide@-SEA, 55, 1-18.

ALVARADO, R. & SELGA, D. (1961). Soil fauna and its agronomic and forestry interest. Revista de la Universidad de Madrid, 10: 451-500.

ARANGO-GALVÁN, A.; CUTZ-POOL, L.; CANO-SANTANA, Z. & LOT, A. (2009).

Community structure of the mulch springtails. Biodiversity of the Pedregal de San Ángel ecosystem. UNAM. Mexico City, 395-402.

ARBEA, J.I. (1988). New species of Odontella (Superodontella) (Collembola, Odontellidae) from Navarra (N Iberian Peninsula). Misc Zool., 12: 109-119.

ARBEA, J.I. & BLASCO-ZUMETA, J. (2001). Ecology of springtails (Hexapoda, Collembola) in Los Monegros (Zaragoza, Spain). Bol. S. E. A., 28: 35-48.

ARBEA, J.I. & JORDANA, R. (1990). Ecology of edaphic springtails populations in a meadow and a pine forest in the sub-Mediterranean region of Navarra. Mediterranea Ser. Biol., 12: 139-148.

ARBEA, J.I.; BAQUERO, E.; BERUETE, E.: PÉREZ FERNÁNDEZ, T. & JORDANA, R.

(2021). Catalogue of the cave springtails of the Ibero-Balearic area and northern Macaronesian islands (Collembola). Bol. S.E.A., 68: 1-80.

ARDANAZ, A. & JORDANA, R. (1983). Contribution to the knowledge of taxonomic characters of Hypogastrura (Ceratophysella) tergilobata Cassagnau, 1954 and H. (Ceratophysella) denticulata (Bagnall, 1941) (Collembola). Proceedings of the I Iberian Congress of Entomology. León. 1: 21-30.

ARROYO, J.; ITURRONDOBEITIA, J.C.; CABALLERO, A.I. & GONZÁLEZ-CARCEDO, S. (2003). An approach to the use of arthropod taxa as bioindicators of soil conditions in agrosystems. Bol. S.E.A., 32: 73-79.

BABENKO, A.; STEBAEVA, S. & TURNBULL, M.S. (2019). An updated checklist of Canadian and Alaskan Collembola. Zootaxa, 4592(1): 1-125.

BARRIENTOS, J.A. (Ed.) (2004). Practical course on Entomology. Servei de publicacions de l'Universitat Autònoma de Barcelona. Barcelona. 947 pp.

BELLINGER, P.F.; CHRISTIANSEN, K.A. & JANSSENS, F. (2003). Checklist of the

Collembola: Families. Department of Biology. University of Antwerp (RUCA). Antwerp, Belgium.

BERUDE, M.; GALOTE, J.K.; PINTO, P.H. & AMARAL, A. (2015). A mesofauna do solo e sua importância como bioindicadora. Enciclopédia Biosfera, 11(22): 14- 28.

BODÍ, M. B.; CERDÀ, A.; MATAIX-SOLERA, J. & DOERR, S. H. (2012). Effects of forest fires on vegetation and soil in the mediterranean basin: literature review. Bulletin of the Association of Spanish Geographers.

BONNET, L.; CASSAGNAU, P. & DEHARVENG, L. (1979). Recherche d'une

Methodology in the analysis of the breakdown of biocenotic equilibria: applications to the edaphitic collemboles of the Pyrenees. Rev. Ecol. Biol. Sol, 16 (3): 373-401.

BURBANO-ORJUELA, H. (2016). Soil and its relationship with ecosystem services and food security. Journal of Agricultural Sciences, 33(2): 117-124.

CERRO, A. DEL & LUCAS, M.E. (2007). The Pinus nigra Arn. In the Serrania de Cuenca: study on its natural regeneration and bases for its management. Forestry series no. 1. Conserjería de medio ambiente y desarrollo rural de la

Junta de Comunidades de Castilla-La Mancha. 56 pp.

CHANDRA, K. (2008). Insecta: Hemiptera. Faunal Diversity of Jabalpur District, MP: 141-157.

CIPOLA, N.G.; DA SILVA, D.D. & BELLINI, B.C. (2018). Class Collembola. In: Hamada, N.; Thorp, J. H.; Rogers, D. C. (Eds.) Thorp and Covich's Freshwater Invertebrates. Vol. 3: Keys to Neotropical Hexapoda 4th ed., pp. 11-55. Academic Press.COLINA, C. L., & ROLDÁN, P. L. (1991). Principal component analysis: application to the analysis of secondary data. Papers: revista de sociologia, 31-63.

CROWSON, R.A. (1981). The Biology of the Coleoptera. Academic press, London- N.Y.-Toronto-Sydney-San Francisco, 802 pp.

CUTZ-POOL, L.Q.; PALACIOS-VARGAS, J.G. & VÁZQUEZ, M. (2003). Comparison of some ecological aspects of Collembola in four plant associations of Noh-Bec, Quintana Roo, Mexico. Folia Entomol. Mex., 42 (1): 91-101.

DALLAI, R. (1973). Ricerche sui collemboli. XVI. Stachorutes dematteisi n. g., n. s., Micranurida intermedia n. s. e considerazionei sul genere Micranurida. Redia, 54: 23-31.

DI CASTRI, F. & VITALI DI CASTRI, V. (1982). Soil fauna of Mediterranean-climate regions. In: Di Castri, F, Goodall, D.W. & Spetcht, R.L. (Eds.). Mediterranean- type shrublands. Elsevier, Amsterdam. 445-478 pp.

DINDAL, D. (1990). Soil biology guide. John Wiley & Sons. New York. 1,349 pp.

DOMÍNGUEZ-FONTANA, C. (2011). Preliminary comparative study of the forests of the Sierra de Juárez, Baja California (Mexico) and the Serranía de Cuenca (Spain) (Doctoral dissertation, Universitat Politècnica de València).

DORAN, J.W.; COLEMAN, D.C.; BEZDICEK, D.F. & STEWART, B.A. (Eds.) (1994).

Defining soil quality for a sustainable environment. Soil Science Society of America, 35. Winsconsin, USA. 35.

EGERT, M.; MARHAN, S.; WAGNER, B.; SCHEU, S. & FREDRICH, M.W. (2004).

Molecular profiling of 16S rRNA genes reveals diet-related differences of microbial communities in soil, gut and casts of Lumbricus terrestris (Oligochaeta: Lumbricidae). FEMS Microbiology Ecology, 48: 187-197.

FJELLBERG, A. (2007). The Collembolan of Fennoscandia and Denmark. Part II: Entomobryomorpha and Symphypleona. Fauna Entomologica Scandinavica, volume 42. Tjöme. 264 pp.

GAMA, M.M.; MURIAS DOS SANTOS, F.A. & NOGUEIRA, A. (1989). Comparison de la composition de populations de Collemboles de peuplements d'eucaliptus (Eucaliptus globulus) et de chêneliège (Quercus suber). In: Dallai, R. (Ed.). 3rd International Seminar on Apterygota. Siena. 345 pp.

GARCÍA-ÁLVAREZ, A. & BELLO, A. (2004). Diversity of soil organisms and transformations of organic matter. Proceedings. I International Conference on Soil and Compost Eco-Biology. León. 211 pp.

GISIN, H.R. (1943). Ökologie und lebensgemeinschaften der collembolen im schweizerischen exkursionsgebiet Basels. Rev. Suisse Zool. 50: 131-224.

GOLDARAZENA, A. (2015). Order Thysanoptera. Ide@-SEA Journal, 52: 1-20.

GONGALSKY, K.B.; MALMSTRÖM, A.; ZAITSEV, A.S.; SHAKHAB, S.V.; BENGTSSON,

J. & PERSSON, T. (2012). Do burned areas recover from inside? An experiment with soil fauna in a heterogeneous landscape. Applied Soil Ecol., 59: 73-86.

GOULA, M. & MATA, L. (2015). Order Hemiptera. Revista Ide@-SEA, 53, 1-30.

HÅGVAR, S. (1998). The relevance of the Rio-Convention on biodiversity to conserving the biodiversity of soils. Applied Soil Ecology, 9: 1-7.

HERRERA, L. (2015). Order Dermaptera. Ide@-SEA Journal, 42: 1-10.

HJORTH-ANDERSEN, M.C.T. (2015). Order Diptera. Ide@-SEA Journal, 63: 1-22.

JOFFE, J.S. (1936). Pedobiology. Rutgers University Press. New Brunswick, New Jersey.

JORDANA, R. & ARBEA, J.I. (1989). Identification key to the genera of Springtails of Spain (Insecta, Collembola). Pub. Biol. Univ. Navarra, Ser. Zool., 19: 1-16.

JORDANA, R.; ARBEA, J.I. & ARIÑO, A.H. (1990). Catalogue of Iberian springtails. Database. Publ. Biol. Univ. Navarra, Ser. Zool., 21: 231.

JORDANA, R.; ARBEA, J.I.; SIMÓN, C. & LUCIÁÑEZ, M.J. (1997). Collembola,

Poduromorpha. In: Fauna Ibérica, vol. 8. RAMOS, M.A. (Ed.) Museo Nacional de Ciencias Naturales. CSIC. Madrid. 807 pp.

LOPES, C.H. & GAMA, M.M. (1994). The effect of fire on collembolan population of Mata de Margaraca (Portugal). Env. J. Soil, 30: 133-141.

LUCIÁÑEZ, M.J. (1990). Contribution to the study of the springtails of the Central Massif of the Sierra de Gredos. Doctoral thesis. Universidad Autónoma de Madrid.

LUCIÁÑEZ, M.J. & INIESTO, P. (2006). Faunistic and ecological study of springtails (Hexapoda, Collembola) communities in burnt pine forests on the southern slopes of the Sierra de Gredos. Boln. Asoc. Esp. Ent., 30 (3-4): 75-95.

LUCIÁÑEZ, M.J. & SIMÓN, J.C. (1987). A study of the soil collembolan population of Raña soils in the province of Guadalajara. Proceedings of the I

Meeting on Soil Biology and Ecology. Pamplona. 417-524.

LUCIÁÑEZ, M.J. & SIMÓN, J.C. (1989). Springtails of the meadows of the Sierra de Gredos (note 1). Boletín del Grupo de Entomológico de Madrid, 4: 5-16.

LUCIÁÑEZ, M.J. & SIMÓN, J.C. (1991). Study of the seasonal variation of the colembofauna in high mountain soils in the Sierra de Guadarrama (Madrid). Misc. Zool., 15: 103-113.

MAGURRAN, A. E. (1988). Ecological diversity and its measurement. Princeton University Press. New Jersey. 179 pp.

MATAIX-SOLERA, J. & GUERRERO, C. (2007). Effects of forest fires on soil properties. Forest fires, soils and water erosion, 5-40.

MATEOS, E. (2008). The European Lepidocyrtus Bourlet, 1839 (Collembola: Entomrobryidae). Zootaxa, 1769(1): 35-59.

MCGAVIN, G.C. (2001). Essential Entomology. Oxford University Press. 350 pp.

MONERO, N.H.; AUTÓNOMO, O.; DE INDUSTRIA, C.; DE COMUNIDADES, J. &

MANCHA, C.L. (2010). Serranía de Cuenca Natural Park. Foresta, (47): 194-196.

MORENO, C.A. (2000). Methods for measuring biodiversity. M&T-Manuales y Tesis SEA, vol. 1. Zaragoza, Spain. 84 pp.

MURILLO-CUEVAS, F.D.; ADAME-GARCIA, J.; CABRERA-MIRELES, H. &

FERNÁNDEZ-VIVEROS, J.A. (2019). Fauna and edaphic microflora associated with different land uses. Ecosistemas y recursos agropecuarios, 6(16): 23-33.

NAVIA, J.F.; BARRIOS, E. & SÁNCHEZ, M. (2006). Effects of aboveground plant biomass inputs on temperature, humidity and soil nematode dynamics during the dry season in Santander de Quilichao (Department of Cauca). Acta agronómica, 55(2): 1-7.

ORTIZ VALBUENA, A. (1992). Contribution to the designation of origin of honey from La Alcarria. Doctoral thesis. Complutense University of Madrid. 311 pp.

PALACIOS-VARGAS, J.G. & MEJÍA, B.E. (2007). Techniques for collection, mounting and preservation of edaphic microarthropods. Ed. National Autonomous University of Mexico.

PARISI, V. (1979). Soil Biology and Ecology. Blume Ecology. Barcelona. 169 pp.

PETERSEN, H. (2002). General aspects of collembolan ecology at the turn of the millennium. Pedobiologia, 46: 246-260.

PLA, L. (2006). Biodiversity: Inference based on the Shannon index and richness. Interscience, 31(8): 583-590.

POTAPOW, A.A.; SEMENINA, E.E.; KOROTKEVICH, A.Y.; KUZNETSOVA, N.A. &

TIUNOV, A.V. (2016). Connecting taxonomy and ecology: Trophic niches of collembolans as related to taxonomic identity and life forms. Soil Biology and Biochemistry, 101: 20-31.

POTAPOW, M. (2001). Synopses on Paleartic Collembola. Vol. 3: Isotomidae. Staatliches Museum für Naturkunde Görlitz. Görlitz, Germany. 605 pp.

PUJADE-VILLAR, J. & FERNÁNDEZ GUYABO, S. (2004). Hymenoptera. In:

BARRIENTOS, J.A. (Ed.). Practical course on Entomology. Servei de publicacions de l'Universitat Autònoma de Barcelona. Barcelona. 813-857 pp.

REGATO, P. & ESCUDERO, A. (1989). Characterisation of Pinus nigra subsp.

Salzmannii communities in the rocky outcrops of the Southern Iberian System. Bot. Complutensis, 15: 149-161.

RESTREPO, L.F. & GONZÁLEZ, J. (2007). From Pearson to Spearman. Revista Colombiana de Ciencias Pecuarias, 20(2): 183-192.

RICHARDS, O.W. & DAVIES, R.G. (1984). Imms. Treatise on Entomology. Volume 2: classification and biology. Ediciones Omega. Barcelona. 998 pp.

RIVAS-MARTÍNEZ, S. (1987). Memoria del mapa de las series de vegetación de España. ICONA, Ministry of Agriculture, Fisheries and Food. Madrid. 268 pp.

RUIZ, M. (1999). Fluctuations in the biodiversity of the edaphic fauna caused by coniferous reforestation in forests of the Central System with special reference to the communities of springtails: ecological and taxonomic study. Doctoral thesis. Autonomous University of Madrid. 576 pp.

RUSEK, J. (1998). Biodiversity of Collembola and their functional role in the ecosystem. Biodiversity & Conservation, 7: 1207-1219.

SALINAS, M. (2007). Regression and correlation models IV: Spearman's correlation. Cienc. Trab, 143-145.

SANJUAN, A.B.; GONZÁLEZ, L. C. & DE MELLO PRADO, R. (2022). Mesofauna edaphic, some studies carried out: Review. INGE CUC, 18(2): 197-208.

SCHELLER, U. (1990). A list of the British Pauropoda with description of a new species of Eurypauropodidae (Myriapoda). Journal of natural history, 24(5): 1179-1195.

SIMÓN, J.C. (1985). Springtails of the savin soils of the province of Segovia. Note I. Graellsia, 31: 213-230.

SOCARRÁS, A. (2013). Edaphic mesofauna: biological indicator of soil quality. Pastures and Forages, 36(1): 5-13.

SUHADI, DHARMAWAN, A.; NAFIAH, K.; AKHSANI, F. & YULIANITA, A. (2012).Comparative study of Collembola community on post fire land, transitional land and control land in Teak Forest Baluran National Park Situbondo. ICoLiST, 1-6.

TEUBEN, A. & SMIDT, G.R. (1992). Soil arthropod numbers and biomass in two pine forests on different soils, related to functional groups. Pedobiologia, 36: 79- 89. TORRADO, M. & BERLANGA, V. (2013). Discriminant analysis using spss. REIRE: revista d'innovació i recerca en educació.
TORRALBA-BURRIAL, A. (2015). Order Embioptera. Ide@-SEA Journal, 44: 1-6. ÚBEDA, X.; MATAIX-SOLERA, J.; FRANCOS, M. & FARGUELL, J. (2021). Great forest fires in Spain and alterations of their regime in the last decades. Geografia, Riscos e Proteção Civil. Homenagem ao Professor doutor Luciano Lourenço, 2: 147-161.

VILLEGAS-GUZMÁN, G.A. & PÉREZ, T.M. (2005). Pseudoscorpions (Arachnida: Pseudoscorpionida) associated with rat nests of the genus Neotoma (Mammalia: Rodentia) from the Mexican highlands. Acta Zoológica Mexicana (new series). 21: 63-77.

VOIGTLÄNDER, K. (2001). Chilopoda - Ecology. In: MINELLI, A. (Ed.). Treatise on Zoology. Anatomy, Taxonomy, Biology. The Myriapoda. Brill. Leiden-Boston. 309 pp.

WRIGHT, J.C. & WEST, P. (2006). Water vapur absorption in the penicillae millipede Polyxenus largus (Diplopoda: Penicillata: Polyxenida): microcalorimetric analysis of uptake kinetics. The Journal of Experimental Biology, 209: 2486-2494. ZANCADA, M.C. & SÁNCHEZ, A. (1994). Role of nematodes in soil biology. Bol. R. Soc. Esp. Hist. Nat. (Sec. Bio.), 91 (1-4): 49-56.

9. ANNEXES

Table 1. Groups of edaphic fauna sampled in the natural forest of Fuentenava de Jábaga. Legend: N, natural; O, autumn; H, leaf litter; S, shallow soil; P, deep soil.

NHO1		NHO2	NHO3	NHO4	NHO6	NSO1	NSO2	NSO3	NSO4	NSO6	NPO1	NPO2	NPO4	NPO5	NPO6	TOTAL
Collembola	37	1	11	4	6	78	46	103	158	61	6	24	265	20	13	833
Acari	785	438	296	376	235	1386	427	900	616	285	443	170	408	30	53	6563
Pseudoscorpionida	0	0	0	0	0	0	0	0	0	0	0	0	0	0	0	0
Coleoptera	0	0	0	0	0	0	0	1	1	0	0	0	0	0	0	2
Larva Coleoptera	0	2	0	3	1	0	2	4	0	6	1	0	0	0	1	12
Dermaptera	0	0	0	0	2	0	0	0	1	0	0	0	0	0	0	3
Diptera	0	0	2	0	0	0	0	0	0	0	0	2	0	0	0	4
Larva Diptera	7	2	10	5	5	0	0	0	1	0	0	3	0	0	0	33
Embioptera	0	0	1	0	0	0	0	0	0	2	0	0	0	0	0	1
Hemiptera	0	0	0	0	0	0	0	0	0	0	0	0	0	0	0	0
Hymenoptera	0	4	0	0	0	0	1	0	0	0	1	0	0	0	0	6
Protura	0	0	0	0	0	0	0	0	0	0	0	0	0	3	0	3
Psocoptera	0	0	0	0	0	0	0	0	2	0	0	0	0	0	0	2
Thysanoptera	1	0	0	0	0	3	1	0	0	0	0	4	4	1	0	14
Chilopoda	0	0	0	0	0	0	0	0	0	0	0	0	0	0	1	1
Pauropoda	0	0	0	0	0	0	1	0	0	0	0	3	9	0	0	10
Symphyla	0	0	0	0	0	1	0	0	0	0	0	6	0	0	1	8
Julida	0	0	0	0	0	0	0	1	0	0	0	0	0	0	0	1
Polyxenide	0	0	0	0	0	0	0	0	0	0	0	0	0	0	0	0
Oligochaeta	0	0	0	0	0	0	0	0	0	0	0	0	0	0	0	0
Nematoda	0	0	0	0	0	0	0	0	1	0	0	0	0	0	0	1
TOTAL	830	447	320	388	249	1468	478	1009	780	354	451	212	686	54	69	7497
Number of groups	4	5	5	4	5	4	6	5	7	4	4	7	4	4	5	17

Table 2. Groups of soil fauna sampled in the burnt forest of Fuentenava de Jábaga.

Legend: I, fire; O, autumn; H, leaf litter; S, shallow soil; P, deep soil.

	ISO 2	ISO 3	ISO 4	ISO 5	ISO 6	IPO 1	IPO 2	IPO 3	IPO 4	IPO 5	IPO 6	TOTAL
Collembola	34	111	7	28	63	20	10	23	7	20	27	350
Acari	119	325	184	122	295	136	65	71	244	122	226	1909
Pseudoscorpionida	0	0	0	0	0	0	0	0	1	0	0	1
Coleoptera	0	0	0	0	0	0	0	0	0	0	0	0
Larva Coleoptera	0	2	2	0	1	1	0	0	0	0	0	6
Dermaptera	0	0	0	0	0	0	0	0	0	0	0	0
Diptera	0	0	0	0	0	0	0	0	0	0	0	0
Larva Diptera	0	0	0	0	0	0	0	0	0	0	1	1
Embioptera	0	0	0	0	0	0	0	0	0	0	0	0
Hemiptera	0	0	0	0	0	0	0	0	1	0	2	3
Hymenoptera	0	0	0	0	0	0	1	0	0	0	0	1
Protura	0	4	0	0	0	0	2	0	1	3	0	10
Psocoptera	0	0	0	0	0	0	0	0	0	0	0	0
Thysanoptera	3	0	0	0	0	0	0	0	0	0	0	3
Chilopoda	0	0	0	0	0	0	0	0	0	0	0	0
Pauropoda	2	7	2	1	4	8	10	3	2	1	19	59
Symphyla	1	3	0	0	0	1	4	3	6	6	2	26
Julida	0	0	0	0	0	0	0	0	0	0	0	0
Polyxenide	0	0	0	0	0	0	0	0	1	0	0	1
Oligochaeta	0	1	1	1	0	0	0	0	0	0	0	3
Nematoda	0	0	1	0	1	0	0	0	0	0	0	2
TOTAL	159	453	197	152	364	166	92	100	263	152	277	2375
Number of groups	5	7	6	4	5	5	6	4	8	5	6	14

Table 3. Groups of edaphic fauna sampled in the forest edge zone of Fuentenava de Jábaga. Legend: B, edge; O, autumn; H, leaf litter; S, surface soil; P, deep soil.

	BHO2	BHO3	BHO4	BSO4	BSO5	Total
Collembola	390	74	252	130	81	927
Acari	163	312	246	198	48	967
Pseudoscorpionida	0	0	0	0	0	0
Coleoptera	0	1	0	0	0	1
Larva Coleoptera	2	0	1	0	1	4
Dermaptera	0	0	0	1	0	1
Diptera	0	0	0	0	1	1
Larva Diptera	4	10	3	1	0	18
Embioptera	0	0	0	0	0	0
Hemiptera	0	0	0	0	0	0
Hymenoptera	0	0	0	1	1	2
Protura	0	0	0	0	0	0
Psocoptera	5	1	5	1	0	12
Thysanoptera	0	1	0	0	1	2
Chilopoda	0	0	0	1	1	2
Pauropoda	0	0	0	0	0	0
Symphyla	0	0	0	0	0	0
Julida	0	0	0	2	1	3
Polyxenide	0	0	0	0	0	0
Oligochaeta	0	0	0	0	0	0
Nematoda	0	0	0	0	0	0
TOTAL	564	399	507	335	135	1940
Number of groups	5	6	5	8	8	12

Table 4. Significant correlations of the edaphic fauna analysis. * indicates that the correlation is significant at the 0.05 confidence level. ** indicates that the correlation is significant at 0.01 confidence level.

SPEARMAN CORRELATION COEFFICIENT			
Couple	Coefficient of correlation	Couple	Coefficient of correlation
Collembola Psocoptera	0,510**	Dermaptera Julidae	0,370*
Acari Coleoptera	0,379*	Hymenoptera Chilopoda	0,369*
Acari Chilopoda	-0,366*	Hymenoptera Julidae	0,378*
Larva_coleoptera Symphyla	-0,388*	Psocoptera Pauropoda	-0,363**
Larva_diptera Psocoptera	0,476**	Chilopoda Julida	0,641**
Larva_diptera Pauropoda	-0,440*	Pseudoscorpionida Polyxenide	1,000**
Diptera Embioptera	0,577**	Pseudoscorpionida Hemiptera	0,670**
Hemiptera Symphyla	0,391*	Pauropoda Symphyla	0,559**
Hemiptera Polyxenide	0,670**	Protura Symphyla	0,501**

Table 5. Springtail species and number of specimens of each collected in the natural area of the Fuentenava de Jábaga forest. Legend: N, natural; O, autumn; H, leaf litter; S, surface soil; P, deep soil.

	NHO1	NHO2	NHO3	NHO4	NHO6	NSO1	NSO2	NSO3	NSO4	NSO6	NPO1	NPO2	NPO4	NPO5	NPO6	TOTAL
Ceratophysella tergilobata	0	0	0	1	1	1	0	1	2	4	0	0	208	0	0	218
Hypogastrura purpurescens	0	0	0	0	0	0	0	0	0	0	0	0	0	0	0	0
Microgastrura duodecimoculata	0	0	0	0	1	0	0	0	0	3	0	0	0	0	0	4
Xenylla schillei	0	0	0	0	2	0	0	0	0	0	0	0	0	0	0	2
Xenyllogastrura octoculata	0	0	0	0	0	0	0	0	0	0	0	0	0	0	0	0
Brachystomella parvula	0	0	0	0	0	0	0	0	0	0	0	0	0	0	0	0
Friesea steineri	0	0	0	0	0	0	0	0	0	0	0	0	0	0	0	0
Bilobella aurantiaca	0	0	0	0	0	0	0	0	0	0	0	0	0	0	0	0
Micranurida pygmaea	0	0	0	1	0	9	0	0	4	0	0	0	0	0	0	14
Pseudachorudina sp.	0	0	0	0	0	0	0	0	0	1	0	0	0	0	0	1
Pseudachorutes parvulus	0	0	0	0	0	0	0	0	0	1	0	0	0	0	0	1
Simonachorutes romeroi	0	0	0	0	0	1	1	0	0	0	0	1	0	0	0	3
Odontellina nivalis	0	0	0	0	0	0	0	0	0	0	0	0	0	0	0	0
Superodontella selgae	0	0	0	0	2	0	0	0	0	0	0	0	0	0	0	2
Protaphorura armata	0	0	0	0	0	0	0	1	0	16	0	0	0	0	0	17
Mesaphorura macrochaeta	2	0	0	0	0	29	17	18	71	15	6	22	47	14	9	250
Wankeliella medialis	0	0	0	0	0	0	0	0	0	0	0	0	3	0	2	5
Entomobrya multifasciata	7	0	0	0	0	5	0	11	0	0	0	0	0	0	0	23
Entomobrya nicoleti	0	0	0	0	0	0	0	0	0	0	0	0	0	0	0	0
Lepidocyrtus lusitanicus	0	0	0	0	0	0	0	0	0	2	0	0	0	0	0	2
Pseudosinella sp.	0	0	0	0	0	0	0	0	0	0	0	0	0	0	0	0
Heteromurus major	0	0	1	0	0	2	0	1	2	0	0	0	0	0	0	6
Ballistura navacerradensis	0	0	0	0	0	0	0	0	0	0	0	0	0	0	0	0
Folsomides	0	0	0	0	0	0	0	0	0	0	0	0	0	0	0	0

parvulus																
Hemisotoma thermophila	0	0	0	0	0	0	0	0	0	0	0	0	0	0	0	0
Isotoma viridis	0	1	2	1	0	11	3	10	4	13	0	0	0	0	0	45
Isotomiella minor	0	0	0	0	0	0	0	0	0	1	0	0	3	5	2	11
Isotomodes bisetosus	0	0	0	0	0	0	0	0	0	0	0	0	0	0	0	0
Isotomurus sp.	0	0	0	0	0	0	0	0	0	0	0	0	0	0	0	0
Parisotoma notabilis	0	0	3	0	0	15	24	57	70	4	0	1	4	1	0	179
Tetrachantella pilosa	24	0	0	1	0	0	0	0	0	0	0	0	0	0	0	25
Megalothorax minimus	0	0	5	0	0	0	0	0	0	0	0	0	0	0	0	5
Sphaeridia pumilis	4	0	0	0	0	5	1	4	5	1	0	0	0	0	0	20
TOTAL	37	1	11	4	6	78	46	103	158	61	6	24	265	20	13	833
Number of species	4	1	4	4	4	9	5	8	7	11	1	3	5	3	3	20

Table 6. Springtail species and number of specimens of each collected in the burnt forest of Fuentenava de Jábaga. Legend: I, fire; O, autumn; H, litter; S, surface soil; P, deep soil.

	ISO 2	ISO 3	ISO 4	ISO 5	ISO 6	IPO 1	IPO 2	IPO 3	IPO 4	IPO 5	IPO 6	TOTAL
Ceratophysella tergilobata	0	3	0	0	1	4	0	0	0	0	0	8
Hypogastrura purpurescens	0	0	0	0	0	0	0	0	0	0	0	0
Microgastrura duodecimoculata	0	0	0	0	0	0	0	0	0	0	0	0
Xenylla schillei	0	0	0	0	0	0	0	0	0	0	0	0
Xenyllogastrura octoculata	5	0	0	4	0	4	0	0	0	0	0	13
Brachystomella parvula	9	0	0	11	0	0	0	0	0	0	0	20
Friesea steineri	0	4	0	0	5	0	0	2	0	0	0	11
Bilobella aurantiaca	0	0	0	0	0	0	0	0	0	0	0	0
Micranurida pygmaea	0	0	0	0	0	0	0	0	0	0	0	0
Pseudachorudina sp.	0	0	0	0	0	0	0	0	0	0	0	0
Pseudachorutes parvulus	2	0	0	0	0	0	0	0	0	0	0	2
Simonachorutes romeroi	0	2	2	0	1	0	0	0	0	0	0	5
Odontellina nivalis	0	0	0	0	1	0	0	2	0	0	0	3
Superodontella selgae	0	0	0	0	0	0	0	0	0	0	0	0
Protaphorura armata	1	0	2	5	25	0	0	1	0	2	2	38
Mesaphorura macrochaeta	2	67	0	4	24	10	2	6	0	0	17	132
Wankeliella medialis	3	26	0	0	3	0	1	6	0	0	3	42
Entomobrya multifasciata	0	1	0	0	0	0	0	0	0	0	0	1
Entomobrya nicoleti	0	0	0	0	0	0	0	0	0	0	0	0
Lepidocyrtus lusitanicus	1	0	0	0	0	0	0	0	1	0	0	2
Pseudosinella sp.	0	0	1	0	0	0	0	0	0	0	0	1
Heteromurus major	0	0	0	0	0	0	0	0	0	0	0	0
Ballistura navacerradensis	0	5	0	0	2	0	0	0	0	0	0	7
Folsomides parvulus	0	0	0	0	0	0	0	0	0	15	3	18
Hemisotoma thermophila	8	0	0	2	0	0	0	0	0	0	0	10
Isotoma viridis	0	0	0	0	0	0	0	0	0	0	0	0
Isotomiella minor	0	0	0	0	0	0	0	3	1	0	0	4
Isotomodes bisetosus	3	0	2	0	1	1	7	1	3	3	2	23
Isotomurus sp.	0	2	0	2	0	1	0	2	2	0	0	9

Parisotoma notabilis	0	0	0	0	0	0	0	0	0	0	0	0
Tetracanthella pilosa	0	0	0	0	0	0	0	0	0	0	0	0
Megalothorax minimus	0	0	0	0	0	0	0	0	0	0	0	0
Sphaeridia pumilis	0	1	0	0	0	0	0	0	0	0	0	1
TOTAL	34	111	7	28	63	20	10	23	7	20	27	350
Number of species	9	9	4	6	9	5	3	8	4	3	5	20

Table 7. Springtail species and number of specimens of each collected in the edge zone of the Fuentenava de Jábaga forest. Legend: B, edge; O, autumn; H, leaf litter; S, shallow soil; P, deep soil.

	BHO2	BHO3	BHO4	BSO4	BSO5	TOTAL
Ceratophysella tergilobata	3	2	0	0	0	5
Hypogastrura purpurescens	4	0	0	0	0	4
Microgastrura duodecimoculata	5	0	0	1	0	6
Xenylla schillei	0	0	0	0	0	0
Xenyllogastrura octoculata	3	0	0	1	0	4
Brachystomella parvula	0	0	0	0	0	0
Friesea steineri	0	0	0	0	0	0
Bilobella aurantiaca	0	1	0	0	0	1
Micranurida pygmaea	0	0	0	0	1	1
Pseudachorudina sp.	0	0	0	0	0	0
Pseudachorutes parvulus	7	1	0	1	2	11
Simonachorutes romeroi	0	0	0	0	0	0
Odontellina nivalis	0	0	0	0	0	0
Superodontella selgae	0	0	0	0	0	0
Protaphorura armata	0	0	0	0	4	4
Mesaphorura macrochaeta	0	0	1	30	33	64
Wankeliella medialis	0	0	0	0	0	0
Entomobrya multifasciata	0	2	1	1	0	4
Entomobrya nicoleti	1	0	0	0	0	1
Lepidocyrtus lusitanicus	0	1	0	0	0	1
Pseudosinella sp.	1	0	0	0	0	1
Heteromurus major	0	1	0	0	3	4
Ballistura navacerradensis	3	0	0	0	0	3
Folsomides parvulus	0	0	0	0	0	0
Hemisotoma thermophila	10	57	0	0	1	68
Isotoma viridis	0	0	0	0	0	0
Isotomiella minor	0	0	0	0	0	0
Isotomodes bisetosus	2	0	0	5	0	7
Isotomurus sp.	0	0	0	0	0	0
Parisotoma notabilis	5	9	0	6	37	57
Tetrachantella pilosa	346	0	250	85	0	681
Megalothorax minimus	0	0	0	0	0	0
Sphaeridia pumilis	0	0	0	0	0	0
TOTAL	390	74	252	130	81	927
Number of species	12	8	3	8	7	19

Table 8. Significant correlations of the springtails analysis. * indicates that the correlation is significant at the 0.05 confidence level. ** indicates that the correlation is significant at the 0.01 confidence level.

SPEARMAN			
Couple	Correlation coefficient	Couple	Correlation coefficient
Ceratophysella tergilobata	0,409*	Xenylla schillei	1,000**
Ballistura navacerradensis		Superodontella selgae	
Hypogastrura purpurescens	0,526**	Xenyllogastrura octoculata	0,449*
Microgastrura duodecimoculata		Pseudachorutes parvulus	
Hypogastrura purpurescens	0,557**	Xenyllogastrura octoculata	0,466**
Xenyllogastrura octoculata		Tetrachantella pilosa	
Hypogastrura purpurescens	0,478**	Xenyllogastrura octoculata	0,557**
Pseudachorutes parvulus		Entomobrya nicoleti	
Hypogastrura purpurescens	0,446*	Xenyllogastrura octoculata	0,358*
Hemisotoma thermophila		Pseudosinella sp.	
Hypogastrura purpurescens	0,557**	Bilobella aurantiaca	0,383*
Ballistura navacerradensis		Pseudachorutes parvulus	
Hypogastrura purpurescens	0,478**	Bilobella aurantiaca	0,478**
Tetrachantella pilosa		Hemisotoma thermophila	
Hypogastrura purpurescens	1,000**	Bilobella aurantiaca	0,456**
Entomobrya nicoleti		Lepidocyrtus lusitanicus	
Hypogastrura purpurescens	0,695**	Brachystomella parvula	0,388*
Pseudosinella sp.		Protaphorura armata	
Microgastrura duodecimoculata	0,438*	Brachystomella parvula	0,572**
Xenylla schillei		Hemisotoma thermophila	
Microgastrura duodecimoculata	0,514**	Friesea steineri	0,424*
Xenyllogastrura octoculata		Simonachorutes romeroi	
Microgastrura duodecimoculata	0,491**	Friesea steineri	0,799**
Pseudachorudina sp.		Odontellina nivalis	
Microgastrura duodecimoculata	0,633**	Friesea steineri	0,640**
Pseudachorutes parvulus		Wankeliella medialis	
Microgastrura duodecimoculata	0,438*	Friesea steineri	0,447*

Superodontella selgae		Isotomurus sp.	
Microgastrura duodecimoculata	0,397*	Friesea steineri	0,651**
Tetrachantella pilosa		Ballistura navacerradensis	
Microgastrura duodecimoculata	0,526**	Micranurida pygmaea	0,417*
Entomobrya nicoleti		Isotoma viridis	
Pseudachorutes parvulus	0,595**	Micranurida pygmaea	0,443*
Hemisotoma thermophila		Heteromurus major	
Pseudachorutes parvulus	0,498**	Micranurida pygmaea	0,444*
Entomobrya nicoleti		Sphaeridia pumilis	
Pseudachorutes parvulus	0,603**	Pseudachorudina sp.	0,383*
Lepidocyrtus lusitanicus		Pseudachorutes parvulus	
Simonachorutes romeroi	0,411*	Pseudachorudina sp.	0,357*
Ballistura navacerradensis		Protaphorura armata	
Mesaphorura macrochaeta	0,361*	Pseudachorudina sp.	0,526**
Wankeliella medialis		Lepidocyrtus lusitanicus	
Mesaphorura macrochaeta	0,501**	Odontellina nivalis	0,489**
Sphaeridia pumilis		Wankeliella medialis	
Protaphorura armata	0,417*	Wankeliella medialis	0,360*
Odontellina nivalis		Ballistura navacerradensis	
Protaphorura armata	0,403*	Folsomides parvulus	0,382*
Folsomides parvulus		Isotomodes bisetosus	
Protaphorura armata	0,372*	Isotoma viridis	0,399*
Isotomodes bisetosus		Isotomurus sp.	
Hemisotoma thermophila	0,446*	Isotoma viridis	0,635**
Ballistura navacerradensis		Sphaeridia pumilis	
Hemisotoma thermophila	0,368*	Isotomodes bisetosus	-0,387*
Lepidocyrtus lusitanicus		Parisotoma notabilis	
Ballistura navacerradensis	0,557**	Isotomodes bisetosus	-0,361*
Entomobrya nicoleti		Sphaeridia pumilis	
Ballistura navacerradensis	0,358*	Parisotoma notabilis	0,698**
Sphaeridia pumilis		Heteromurus major	
Tetrachantella pilosa	0,357*	Parisotoma notabilis	0,461**
Entomobrya multifasciata		Sphaeridia pumilis	
Tetrachantella pilosa	0,478**	Entomobrya multifasciata	0,509**
Entomobrya nicoleti		Sphaeridia pumilis	
Heteromurus major	0,394*	Entomobrya nicoleti	0,695**
Sphaeridia pumilis		Pseudosinella sp.	

Printed by Books on Demand GmbH, Norderstedt / Germany